ISBN 979-8-9944770-4-5
Printed by Book Logix
First Edition, 2025
Nerdy Brain, LLC.
6631 Princeton Park Ct
Lithonia, GA 30058

Second Edition, 2026
VIQ Enterprises, LLC.

Book Cover by Hiroart
Illustrations by Hiroart

Meet Your Authors

Valerie Bennett, Ph.D., Ed.D.

Dr. Valerie Bennett received a B.E. Degree in Mechanical Engineering from Vanderbilt University, a PhD in Mechanical Engineering from the Georgia Institute of Technology, and an EdD in Higher Educational Leadership from Clark Atlanta University. As a scholar, researcher, and authority on STEM Equity and Curriculum and Program Design, she has led teams to exceed school and state standards by providing Professional Development & Training on STEM Integration and culturally relevant pedagogical practices. Among her accolades, she was selected as a Smithsonian Science Mentor, UNCF Faculty Fellow, and was recognized by the White House Initiative as an HBCU Scholar. She is co-author of the blog "Chalkboards to AI" for the National Science Teacher Association. Her innovative applications for using robotics, coding, drones, design thinking, and artificial intelligence contribute to her research in STEM equity. With her 25 years of experience in the STEM industry and author of 7 books, as a professor, coach, and author, Dr. Bennett strives to ensure that there is more representation in STEM in corporate, universities, and to make sure that our classrooms are equitable. She is intentional with setting kids up on the right track to succeed and thrive in STEM.

Natoshia Anderson, Ed.D.

Dr. Natoshia Anderson is a powerhouse in STEM education, equity, and advocacy, with a career that bridges mechanical engineering, curriculum design, and transformative leadership. As Founder and CEO of The Anderson Strategy Group, she helps schools, nonprofits, and companies design scalable, equitable STEM programs that prepare students for thriving futures. A former Mechanical Engineer turned educator, she has written award-winning curriculum, trained thousands of educators, and impacted more than 100,000 students worldwide. She has secured over $1.85 million in funding for schools and nonprofits and hosts STEMming in Stilettos, a global podcast that amplifies the voices of women of color in STEM.

As an author, Dr. Anderson has published children's books like ABC Engineering and London Gets Curious, co-authored Triumph in the Trenches: Navigating Success for Black Professionals, and, most recently, contributed to Black in the Middle: Navigating Power, Politics, and Purpose in Higher Education.

A
is for
Algorithm

A is for Algorithm

What is an Algorithm?

An algorithm is like a recipe! AI follows step-by-step instructions to solve problems, just like you follow steps to bake cookies or brush your teeth.

Let's Dive Deeper

Create an algorithm for brushing your teeth. Write out steps or act them out in order.

Example: Step One: Grab your toothbrush, Step Two: Run it under the water. What comes next?

This step by step process creates an algorithm that helps AI do its best.

Laughing Corner

Why did the AI chef get an award?
Because its recipe was a-log-rhythm-ic!

B is for Bot

B is for Bot

What is a Bot?

A bot is a friendly robot that helps with tasks! Some bots clean, some play games, and some even tell jokes.

Let's Dive Deeper

Draw your own AI helper bot!
What does it do? What does it look like?

Bots can talk, bots can play,
They help us learn every day!

Laughing Corner

Why don't robots get lost? Because they always follow the right path!

python
# Ask the user to input the fi
word1 = input("Enter the first
# Ask the user to input the se
word2 = input("Enter the second
# Print each word individually
print(word1)
print(word2)
# Combine and print the full se
full_sentence = word1 + " " + wo
print(full_sentence)

C
is for
Code

C is for Code

What is Code?

Code is a special language that tells computers what to do! It's how we make apps, games, and websites.

Let's Dive Deeper

Help your friend get across the room by creating a simple code using arrows to guide them.

Code is magic, code is fun,
It tells computers how to run!

Laughing Corner

What's a robot's favorite snack? Micro-chips!

D is for Data

D is for Data

What is Data?

AI learns from data, just like we learn from books! It looks at lots of information to get smarter.

Let's Dive Deeper

Take different objects and sort them into groups - this showcases how AI sorts data.

AI looks at data, big and small,
It learns and grows—so smart, so tall!

Laughing Corner

Why did the computer go to school? To improve its data-base!

E
is for
Engine

E is for Engine

What is an AI Engine?

An AI engine is like a brain for computers! It helps them think, solve problems, and make smart choices.

Let's Dive Deeper

Solve a puzzle—just like AI solves problems!

An engine helps AI think real fast,
It finds the answers that will last!

Laughing Corner

Why did the robot blush? Because it had a mega-bite!

F
is for
Face
Recognition

F is for Face Recognition

What is Face Recognition?

AI can recognize faces, like how you know your friends!

Let’s Dive Deeper

Play a "Guess Who?" game with pictures of friends or family.

AI sees faces, smiles, and eyes,
It helps your phone—what a surprise!

Laughing Corner

Why did the AI look in the mirror? To recognize itself!

G is for Game AI

G is for Game AI

What is Game AI?

AI makes video games fun by creating smart characters!

Let's Dive Deeper

Play a game and notice how AI characters move!

AI in games, so fun to play,
It makes the bad guys run away!

Laughing Corner

Why did the AI lose at chess?
Because it needed to reboot!

H
is for
Hardware

H is for Hardware

What is Hardware?

Hardware is the body of a computer or robot. It's the parts you can touch, like the keyboard, screen, and even robot arms!

Let's Dive Deeper

Find different hardware in the classroom! Point to the keyboard, screen, and mouse.

Hardware is the parts you see,
A robot's arms, its legs, its feet!

Laughing Corner

Why did the robot go on a diet?
Because it had too many chips!

I
is for Input

I is for Input

What is Input?

Input is information that goes into a computer or AI. When you speak to a voice assistant or tap a button, that's input!

Let's Dive Deeper

Play "Simon Says" but call it "AI Says!" Only follow commands when "AI Says" first.

When you press a key or speak out loud,
That's an input—make AI proud!

Laughing Corner

What do you call an AI that sings?
A-melody-gorithm!

J
is for
Jobs

J is for Jobs

What are Jobs?

AI helps people with jobs, like driving cars, answering questions, and even helping doctors. But people still do the thinking and big decisions!

Let's Dive Deeper

Play "AI Job Match" – match AI tools to jobs (like self-driving cars for drivers).

AI helps in many ways,
But humans lead—it's here to stay!

Laughing Corner

Why did the AI become a chef?
Because it had great "byte"-s of food!

K
is for
Knowledge

K is for Knowledge

What is Knowledge?

AI learns by gathering knowledge, just like you learn in school. It remembers patterns and helps answer questions.

Let's Dive Deeper

Play "AI Learns Too!" – Give AI (a friend playing AI) a fun fact and see if they remember it later.

AI learns like me and you,
The more it knows, the more it can do!

Laughing Corner

What does an AI say when it gets something wrong?
"I need to re-calculate that!"

Hello World!
¡Hola, Mundo!
Bonjour, le Monde !
L
is for
Language

L is for Language

What is Language?

AI can understand and use language, just like humans! It helps computers read, write, and even talk back to us. Think of chatbots or translation apps.

Let's Dive Deeper

Play "Robot Translator." Say simple words ("cat," "dog," "happy") and have the kids "translate" them into robot sounds (beeps, boops) or another language they know.

Language is words, both written and heard, AI can learn them, every single word!

Laughing Corner

Why did the AI bring a dictionary to the party?
To help define the fun!

M is for Machine Learning

M is for Machine Learning

What is Machine Learning?

Machine learning is how AI gets smarter over time. It looks at lots of data and figures things out on its own!

Let's Dive Deeper

Show a set of animal pictures and let kids guess which animal AI would choose as a pattern.

Machine learning helps AI grow,
It watches, learns, and starts to know!

Laughing Corner

What did the AI say after learning something new?
"I'm a smart cookie now!"

N is for Neural Network

N is for Neural Network

What is a Neural Network?

AI has something like a brain called a neural network. It helps AI recognize faces, voices, and even understand what you say!

Let's Dive Deeper

Pretend to be neurons! One kid starts with a message, and others pass it along.

Neural networks work like brains,
Sending signals, like little trains!

Laughing Corner

Why did the AI go to the gym?
To strengthen its neural "nets!"

O
is for
Optimizing

O is for Optimizing

What is Optimizing?

AI helps make things better, or "optimize" them. It finds the best way to do things, like choosing the fastest way to get to school!

Let's Dive Deeper

Draw different paths on a board and let kids find the fastest route.

AI helps to make things right,
Choosing best and saving time!

Laughing Corner

Why did the AI cross the road?
To optimize the best route!

P
is for
Program

P is for Program

What is a Program?

A program is a set of instructions that tells AI what to do, like a robot set you put together.

Let's Dive Deeper

Write simple "programs" for each other (e.g., "Step 1: Clap, Step 2: Jump").

A program tells AI what to do,
Like putting together a robot kit, baking cookies, or any step by step instructions.

Laughing Corner

Why did the AI bring a ladder?
Because it was learning to "code" climb!

Q is for Questions

Q is for Questions

What is a Question?

AI loves questions! When you ask a smart speaker or chatbot something, it finds the answer for you.

Let's Dive Deeper

Play "Ask AI" – Kids take turns asking fun questions and finding answers.

Ask a question, AI might know,
It finds answers fast, not slow!

Laughing Corner

Why did the AI start a Q&A club?
Because it had all the "byte"-sized answers!

R is for Robots

R is for Robots

What is a Robot?

Robots use AI to think and move. Some robots help build cars, and some can even dance!

Let's Dive Deeper

Play "Robot Dance" – Kids pretend to be robots and move only when given commands.

Some robots build, some robots play,
With AI's help, they work all day!

Laughing Corner

Why did the robot go to music school?
To learn some "beep"-boxing!

S
is for
Smart
Devices

S is for Smart Devices

What is a Smart Device?

Smart devices, like phones and tablets, use AI to help you talk to friends, play music, and even turn off the lights!

Let's Dive Deeper

Show pictures of devices and play "Smart or Not?" Kids guess which ones are smart.

Smart devices hear your call,
They turn on lights or play some ball!

Laughing Corner

Why was the smartphone so good at math?
It had lots of "smart" functions!

T is for Training

T is for Training

What is Training?

AI needs training to learn new things, just like you practice math or reading. The more it learns, the smarter it gets!

Let's Dive Deeper

"Train the Robot" – Kids give commands like "clap, jump, spin" in order and the "robot" follows them.

Training helps AI to grow,
Practice makes it smart, you know!

Laughing Corner

What did the AI say after homework?
"That was some hard training data!"

U
is for
User

U is for User

What is a User?

The 'User' is the person using the AI – that's YOU! AI tools and apps are built to help users like you learn new things, finish tasks, or just have fun.

Let's Dive Deeper

"Design Your Helper Bot!" Ask kids to draw or describe their perfect AI helper. What would it look like? What helpful things would the user want it to do (e.g., help with homework, tell jokes, play music, find lost socks)?

U is for User, that means you!
AI is waiting to help you through!

Laughing Corner

What do you call people who use AI all the time?
The user-al suspects!

V
is for
Voice
Assistant

V is for Voice Assistant

What is a Voice Assistant?

A voice assistant is AI that listens when you talk and helps you with answers, reminders, and even jokes!

Let's Dive Deeper

Take turns playing voice assistant—one kid asks a question, another answers using a silly "AI voice."

Say 'Hello' and it replies,
A voice assistant's super wise!

Laughing Corner

Why did the voice assistant bring a map?
Because it didn't want to lose its "directions!"

W
is for
Website

W is for Website

What is a Website?

Websites use AI to help you find information fast—like when you look for pictures, songs, or facts!

Let's Dive Deeper

Let kids imagine their own website and draw what it would have on it.

Click a link, take a ride,
With AI, the web's your guide!

Laughing Corner

What kind of website does a robot like?
A site with a good "interface!"

X is for XAI

X is for XAI

What is XAI?

X is for XAI! That stands for Xplainable Artificial Intelligence. It's AI that can show its thinking steps, explaining how it got an answer or why it made a decision, so we can understand it better and trust it!

Let's Dive Deeper

"Show Your Steps!" Give kids a simple task, like sorting three different colored blocks or drawing a path through a simple maze on paper. Ask them to do the task and then explain the steps they took to finish it ("First I picked up the red block, then the blue...").

XAI, XAI, clear and bright,
Shows how AI gets things right!

Laughing Corner

Why is XAI so trustworthy?
Because it always shows its work!

Y
is for
You

Y is for You

What about You?

Yes, YOU can create AI too! Maybe you'll design a smart robot, build an AI-powered game, or teach AI something new one day.

Let's Dive Deeper

Imagine you could invent your own AI. What would it do? Draw and describe it!

You can code, you can create,
With AI, the future's great!

Laughing Corner

Why did the AI become a teacher?
Because it loved helping students compute!

Z
is for
Zoom

Z is for Zoom

What is Zoom?

AI helps with video calls, making it easier to see and talk to people from anywhere in the world!

Let's Dive Deeper

Pretend you're on a video call with a robot—what would you say?

Zoom connects us, near and far,
AI helps—it's the tech star!

Laughing Corner

What did the computer say to the webcam?
Stop staring—I'm working!

www.ingramcontent.com/pod-product-compliance
Lightning Source LLC
LaVergne TN
LVHW060044210926
848481LV00013B/3954